SMALL BUT FIERCE

John Allan

Picture Credits
(abbreviations: t = top; b = bottom; m = middle; l = left; r = right; bg = background)

Shutterstock: 24-25bg, 30tl; Aaron J Seltzer 29tr; Alexander Wong 12-13bg; Arief Budi Kusuma 28br; Charly Morlock 30ml; Danita Delimont 28tr; Dirk Ercken 2tl, 7mr, 20-21bg, 31mr; Eric Isselee 4-5bg; Griffin Gillespie FC, 31bl; Jeremy1234 29br; karegg 7tl; Kessler Bowman 8-9bg; Lubos Houska 6bl, 31tl; Michael Lynch 16-17bg; Pavel Krasensky 6mr; Pitiya Phinjongsakundit 26-27bg, 28mr; Protasov AN 22-23bg; shaftinaction 3b, 18-19bg; torook 1bg, 14-15bg; Vladimir Turkenich 10-11bg; Young Swee Ming 30br.

Every effort has been made to trace the copyright holders and we apologise in advance for any unintentional omissions. We would be pleased to insert the appropriate credit in any subsequent edition of this publication.

Copyright © 2025 Hungry Tomato Ltd

First published in 2025 by Hungry Tomato Ltd
F15, Old Bakery Studios, Blewetts Wharf, Malpas Road, Truro, Cornwall, TR1 1QH, UK.

No part of this publication may be reproduced, stored in a retrieval system, or transmitted in any form or by any means, electronic, mechanical, photocopying, recording, or otherwise, without prior written permission of the copyright owner.
A CIP catalogue record for this book is available from the British Library.

ISBN 9781835690772

Printed in China

Discover more at
www.hungrytomato.com

SMALL BUT FIERCE

MEET SOME OF THE WORLD'S MOST DANGEROUS ANIMALS!

CONTENTS

Small But Fierce	6	Poison Dart Frog	20
Harvester Ant	8	Deathstalker	22
Pangolin	10	Fierce Snake	24
Western Diamondback Rattlesnake	12	Mosquito	26
Tarantula	14	When Predators Become Prey	28
Vampire Bat	16	Fearsome Facts	30
Funnel-Web Spider	18	Glossary & Index	32

Words in **BOLD** can be found in the glossary.

SMALL BUT FIERCE

Which animal wins the title of the most fierce creature? This big question isn't as easy to answer as you might think! There's lots to consider…

THE WORLD'S SMALLEST BUT DEADLIEST

We have included the smallest but most dangerous creatures on Earth within the pages of this book…

SMALL BUT MIGHTY

The deadliest animals come in all shapes and sizes. Explore the top ten, small but fierce creatures, from venomous snakes to blood-sucking mosquitos…

HUNGRY PREDATORS

Carnivores are animals that only eat meat, herbivores are animals that only eat plants, and omnivores are animals that eat both! Most of the extreme **predators** we explore in this book are carnivores, which makes them the most dangerous of all!

DEADLY COUNTDOWN

All animals in this book have been ranked in order, from the 10th most deadly hunter down to the ultimate predator. It's not always the biggest animals that win!

WARNING

THINGS GET GRIM FROM HERE ON IN... TURN THE PAGES TO FIND OUT MORE!

HARVESTER ANT

They may be little, but you don't want to get in the way of harvester ants! If they feel threatened, they have a sharp **sting** and painful **venom**, which is strong enough to harm animals much bigger than themselves!

A TWO-PRONGED ATTACK
With a ferocious bite, harvester ants pin their **prey** down and use their pointy stinger to inject a deadly venom. Ouch!

ANYBODY HOME?

Harvester ants live in huge nests, called mounds, that stretch up to 1 metre (3 ft) underground in a series of deep tunnels. They are protective of their **territory**, and have been known to kill animals that disturb their nests!

FACT FILE

WEIGHT
Up to 5 mg (1 lb)

DIET
Omnivore

LOCATION
United States

LETHAL POWERS
Powerful sting, strong body, and deadly venom

DEADLY COUNTDOWN

NO.10

PANGOLIN

The pangolin may not look fierce, but it's a mighty hunter! It uses its powerful claws to burrow into termite mounds and ant nests, and then uses its long tongue to slurp up their eggs. Its hard, scale-covered coat protects the pangolin against bites and stings.

ANTS BEWARE!
A pangolin's long sticky tongue is the perfect hunting tool and can be longer than their whole body! This means they can reach deep down inside an ants nest and catch them unaware. Slurp!

FACT FILE

WEIGHT
Up to 33 kg
(73 lbs)

DIET
Carnivore

LOCATION
Africa and Asia

LETHAL POWERS
Large claws, long tongue, and sharp scales

DEADLY COUNTDOWN

NO.9

WALKING PINECONES
A pangolin's **scales** are very unusual for a **mammal**. They are made of **keratin**, which makes them really tough! The super sharp scales on a pangolin's tail can be used to lash out against attackers.

11

WESTERN DIAMONDBACK RATTLESNAKE

The western diamondback rattlesnake is known for its rattle tail, which helps warn off any danger - and for good reason! This lethal killer bites fast and deep, injecting deadly venom into its victims.

SNAKE SENSES

Western diamondbacks belong to a group of snakes called pit vipers. This means they have a heat sensing pit in each nostril which can detect changes in temperature. This is a helpful tool for tracking prey!

ARMED AND DANGEROUS

This snake has one of the deadliest bites in the world! They can slither up and attack prey very fast. Their strong, curved fangs hold onto prey tightly so they can't escape.

FACT FILE

WEIGHT
Up to 7 kg
(15 lbs)

DIET
Carnivore

LOCATION
North America

LETHAL POWERS
Agile, deadly venom, and sharp fangs

DEADLY COUNTDOWN

NO.8

TARANTULA

Tarantulas are one of the largest spiders in the world! Unlike other spiders, tarantulas don't spin webs. Instead, they're active hunters that prowl around at night looking for prey. They mostly eat insects, but they sometimes target bigger animals such as frogs and mice!

FANG-TASTIC
Once a tarantula has caught their prey, they use their fangs to bite into it. This injects a venom that turns their prey into liquid, which they suck up through their straw-like mouth. Yuck!

FACT FILE

WEIGHT
Up to 85 g (3 oz)

DIET
Carnivore

LOCATION
Africa, Asia, Australia, and America

LETHAL POWERS
Sharp fangs, **paralysing** venom, and sneaky trip wires

DEADLY COUNTDOWN

NO.7

NOCTURNAL NIGHTMARES
Despite having eight eyes, tarantulas actually have very bad eyesight! This means they have to rely on other senses. Tarantulas are very sensitive to sound, so they set up trip wires outside their burrows in order to hear prey.

VAMPIRE BAT

This mammal has a fearsome reputation because of its diet - it feeds on the blood of other mammals! Unlike the mythical beast it's named after, the vampire bat does not actually suck blood – it bites its victims, then laps up the blood with its tongue.

LIGHT BUT LETHAL
Vampire bats are so light that they can drink the blood of an animal for more than 30 minutes without waking it up!

FACT FILE

WEIGHT
Up to 55 g
(2 oz)

DIET
Carnivore

LOCATION
America

LETHAL POWERS
Agile, sharp fangs, and strong senses

DEADLY COUNTDOWN

NO.6

BLOOD SUCKERS!
Heat sensors on a vampire bat's nose help them find a good spot to feed from. In fact, their senses are so strong that they can remember breathing patterns, and return to the same animal night after night!

FUNNEL-WEB SPIDER

The funnel-web spider is one of the most dangerous spiders on Earth. Although it mainly lives in woodland, the funnel-web spider can also be found underneath houses and in garages.

STICKY HUNTER
The funnel-web spider hunts by making a web in the shape of a funnel and waiting for its prey to get stuck in it. Once their prey is caught, this spider releases its deadly venom!

FACT FILE

WEIGHT
Unknown

DIET
Carnivore

LOCATION
Australia

LETHAL POWERS
Deadly venom, agile, and intelligent hunter

DEADLY VENOM
A funnel-web spider can kill prey in 15 minutes with just one bite. Unlike most other spiders, the funnel-web spider can be very deadly to humans as well as animals!

DEADLY COUNTDOWN

NO.5

POISON DART FROG

Most people think that small frogs are harmless, but the poison dart frog would prove them wrong! This frog lives in tropical rainforests, where its variety of markings – yellows, oranges, reds, greens, and blues – warn predators not to eat them!

FIERCE DEFENDERS

Poison dart frogs use the **poison** they make to defend themselves from other predators. Their skin is covered in a poisonous slime, so just touching a poison dart frog can kill prey, no matter the size!

DESTRUCTIVE DIET

This frog mainly feeds on insects, especially ants, which it needs to eat in order to produce its poison! This poison is then released from its body onto its skin, ready to kill its prey!

FACT FILE

WEIGHT
Up to 30 g (1 oz)

DIET
Carnivore

LOCATION
Central and South America

LETHAL POWERS
Poisonous slime, incredible agility, and bright markings

DEADLY COUNTDOWN

NO.4

DEATHSTALKER

The deathstalker is the most dangerous scorpion in the world! Found in the deserts and scrubland of Africa, this scorpion is feared by the local people that live there. The deathstalker is also known as the "Palestine yellow scorpion".

DANGEROUS PINCERS
The deathstalker hunts insects, but doesn't use its sting. Instead, it tears its prey apart with its **pincers**!

VENOMOUS TAIL
Deathstalkers only use their sharp tail to inject deadly venom when they feel threatened. This scorpion will often sting its victim over and over again!

FACT FILE

WEIGHT
Up to 3 g
(0.1 oz)

DIET
Carnivore

LOCATION
Middle East and
North Africa

LETHAL POWERS
Strong pincers, sharp tail,
and deadly venom

DEADLY COUNTDOWN

NO.3

FIERCE SNAKE

The fierce snake, also known as the inland taipan, is not as famous as cobras and rattlesnakes, but it's much more dangerous! It has the deadliest venom of any snake on land. Luckily for its prey, the fierce snake is very rare!

SNEAKY HUNTER

Fierce snakes can strike faster than the eye can see! They mainly hunt small mammals, which they swallow whole. As they bite their victims, two fangs inject a small dose of venom. Prey normally dies within seconds.

FACT FILE

WEIGHT
Up to 2 kg
(4 lbs)

DIET
Carnivore

LOCATION
Australia

LETHAL POWERS
Sharp fangs, deadly venom, and agile

REMAIN CALM
As well as being rare to find, the fierce snake is usually very shy. If you are unlucky enough to see one, slowly walk away, because they can be very aggressive if disturbed...

DEADLY COUNTDOWN

NO.2

MOSQUITO

'Mosquito' means 'little fly', but don't be fooled by their small size. Mosquitoes drink human blood! Some mosquitoes drink human blood, and can pass on a deadly disease called **malaria**. This disease has killed more people than any other on our planet.

TASTY BLOOD
Only female mosquitoes drink blood, as it helps them to lay eggs. They are only interested in drinking the blood of mammals – humans are easy targets! Mosquito adults also feed on fruits for energy, so they can fly for longer and find as much prey as possible!

FATAL SUCKERS

Mosquitoes do not have teeth. Instead, they have feeding tubes that look like sharp needles! They jab these tubes into human skin to suck up blood. Mosquitoes only need to take a tiny drop of blood for it to be deadly to humans!

FACT FILE

WEIGHT
Under 3 mg (7 lbs)

DIET
Omnivore

LOCATION
Nearly everywhere!

LETHAL POWERS
Agile, sharp needle-like mouth, and carries diseases

DEADLY COUNTDOWN

NO.1

WHEN PREDATORS BECOME PREY

These tiny but mighty predators are some of the most fierce in the animal kingdom. But what happens when these deadly hunters are hunted themselves?

WESTERN DIAMONDBACK RATTLESNAKE

Even with their deadly venom, western diamondback rattlesnakes are prey to some animals. Coyotes, birds of prey, and even other snakes eat this aggressive hunter for breakfast!

PANGOLIN

Pangolins face many threats, and are close to **extinction** as a result. Loss of their natural habitat and illegal hunting has meant that laws have been put in place to try to protect them.

TARANTULA

While lethal hunters themselves, tarantulas have many predators. The tarantula hawk is a type of wasp that specifically preys on these deadly spiders!

POISON DART FROG

Poison dart frogs are endangered animals. Every year, their numbers decrease due to climate change and the destruction of their rainforest habitat.

FEARSOME FACTS

There are so many deadly and fearsome facts about each tiny but terrifying predator. Here are some more that show just how impressive these predators are...

TARANTULA
Some tarantulas can live for up to 30 years – that's old for a spider!

HARVESTER ANT
Harvester ants need a queen harvester ant to keep their colony going. Without a queen harvester ant, the others would die out after a few weeks!

MOSQUITO
Mosquitoes can drink their entire body weight in blood during one meal!

PANGOLIN
The pangolin doesn't have teeth, so everything it eats is swallowed whole!

FIERCE SNAKE
These snakes lay eggs! They can produce up to 20 eggs per breeding season.

FUNNEL-WEB SPIDER
Although this spider's bite can be fatal for humans, other mammals, like cats and dogs, are not affected as badly.

POISON DART FROG
Native people sometimes put this frog's poison on their arrows when they go hunting..

WESTERN DIAMONDBACK RATTLESNAKE
These rattlesnakes are born with a tiny button on the end of their tail. A new rattle forms each time the snake sheds its skin!

VAMPIRE BAT
Unlike other types of bats, vampire bats move in all sorts of ways to catch their prey. They walk, run, and jump, as well as fly!

DEATHSTALKER
Scorpions belong to a group of animals called arachnids, the same group as spiders. They have eight legs like spiders, too!

GLOSSARY

Agile - able to move quickly and easily.

Extinction - when something like a plant or animal species no longer exists.

Keratin - a hard substance that makes up part of an animal's body and your hair. Our fingernails are made of this.

Malaria - an infectious disease transmitted to humans by the bite of an infected mosquito.

Mammal - a warm-blooded animal with a covering of hair on the skin and the ability to produce milk to feed their young.

Nocturnal - an animal that is active at night and sleeps during the day.

Paralysing - unable to more all or part of the body.

Pincers - the front claws of a lobster, crab, scorpion, or similar creature.

Poison - a substance that can kill or hurt a person or animal.

Predators - animals that live by attacking and killing other animals.

Prey - an animal hunted or caught for food.

Sting - a sharp, piercing part of an animal, often ejecting a venomous substance.

Scales - hard plates that protect an animal's skin.

Territory - a specific area that belongs to or is controlled by someone or something.

Venom - a poisonous substance of an animal, usually passed on by a bite or a sting.

INDEX

C
Carnivore (meat-eater) 7, 11, 13, 15, 17, 29, 21, 23, 25

F
Fierce snake 24-25, 31
Funnel-web spider 18-19, 31

H
Harvester ant 8-9, 30
herbivore (plant-eater) 7

M
Mosquito 6, 26-27, 30

O
omnivore 7, 9, 27

P
Pangolin 10-11, 28, 30
Deathstalker 22-23, 31
Poison dart frog 20-21, 29, 31

T
Tarantula 14-15, 29, 31

W
Western diamondback rattlesnake 12-13, 28, 30

V
Vampire bat 10-11, 31